TRAINING AI

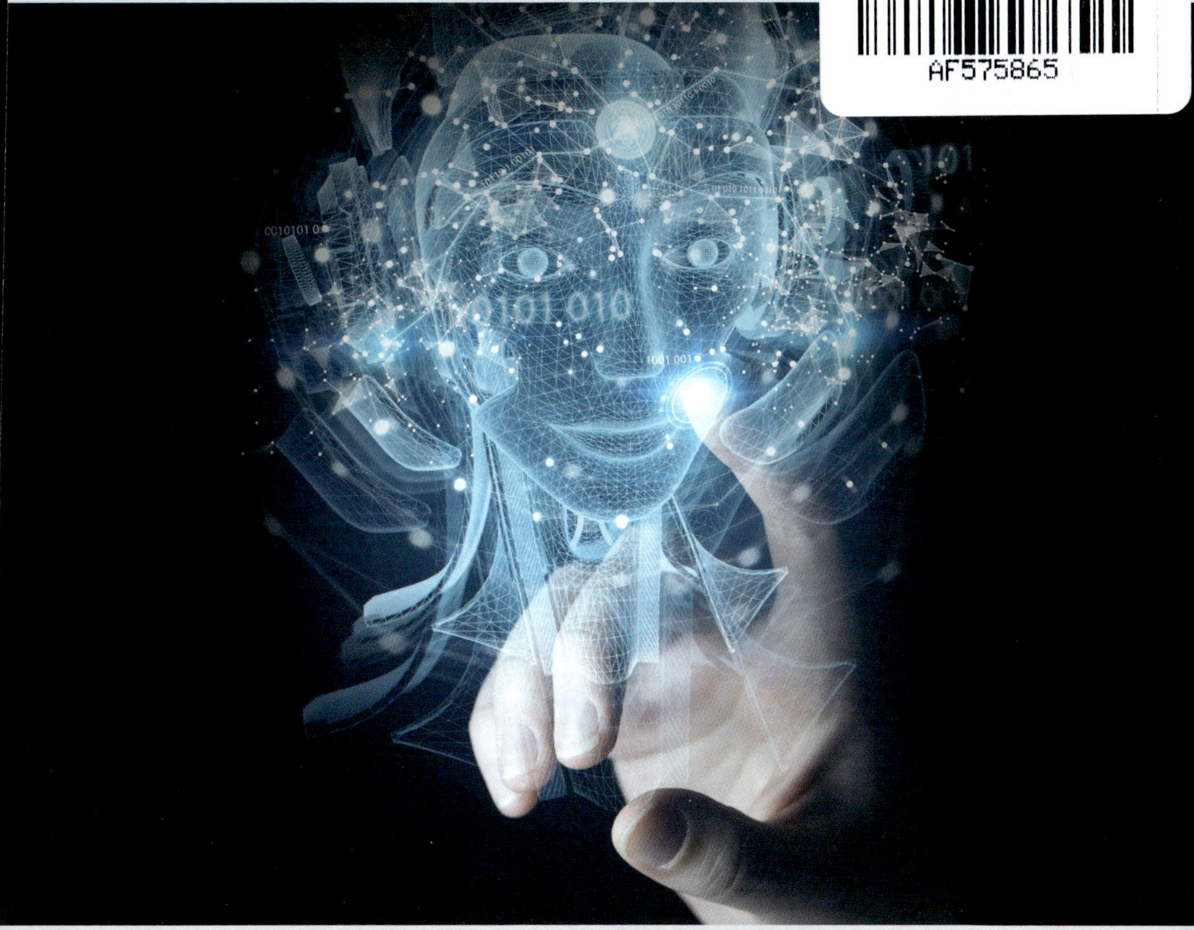

by Josh Gregory

Cherry Lake Press
Ann Arbor, Michigan

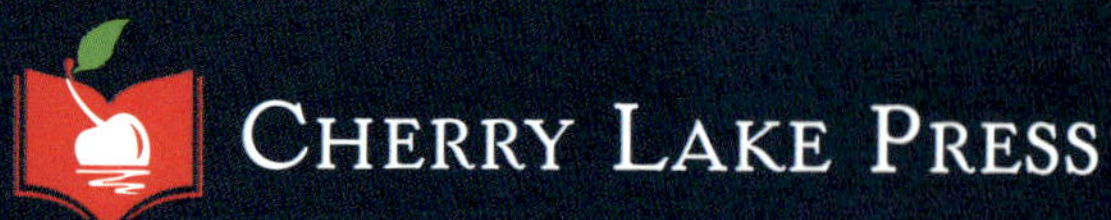

Published in the United States of America by Cherry Lake Publishing
Ann Arbor, Michigan
www.cherrylakepublishing.com

Reading Adviser: Beth Walker Gambro, MS, Ed., Reading Consultant, Yorkville, IL

Photo Credits: ©sdecoret/Shutterstock, cover, title page; © Iftikhar Alam/Dreamstime.com, 5; © Aoleshko/Dreamstime.com, 6; © Dtiberio/Dreamstime.com, 9; © Adrianferrerfilms/Dreamstime.com, 11; © Waingro/Dreamstime.com, 13; © Chekyravaa/Dreamstime.com, 14; © Shao-chun Wang/Dreamstime.com, 17; © Phuttharak Chindarot/Dreamstime.com, 19; © Yuri Arcurs/Dreamstime.com, 21; © Pop Nukoonrat/Dreamstime.com, 25; © Iurii Motov/Shutterstock, 26; © Waingro/Dreamstime.com, 27; © TatiVovchenko/Shutterstock, 28; © Serhii Akimov/Dreamstime.com, 29

Cherry Lake Press is an imprint of Cherry Lake Publishing Group.

Library of Congress Cataloging-in-Publication Data has been filed and is available at catalog.loc.gov.

Cherry Lake Press would like to acknowledge the work of the Partnership for 21st Century Learning, a Network of Battelle for Kids. Please visit Battelle for Kids online for more information.

Printed in the United States of America

ABOUT THE AUTHOR

Josh Gregory is the author of more than 200 books for kids. He has written about everything from animals to technology to history. A graduate of the University of Missouri–Columbia, he currently lives in Chicago, Illinois.

CONTENTS

Chapter 1

COMPUTERS WITH MINDS OF THEIR OWN

It seems like everyone is talking about artificial intelligence, or AI. It's all over the news. New AI technology is being created all the time. People are finding new and creative ways to use it. But what is AI?

AI might make you think of the robots in a movie like *Star Wars*. They walk and talk like people. Or you might think of a computer that answers any question you ask. These things were once imaginary. They were only in science fiction books and movies. Today, they are closer to reality than ever before. Some modern AI systems are even more advanced than the ones people once imagined.

Different people have different definitions of AI. Most agree that it has to do with computer systems that work like the human mind. This means the systems can study **data** and learn from it. They can use the things

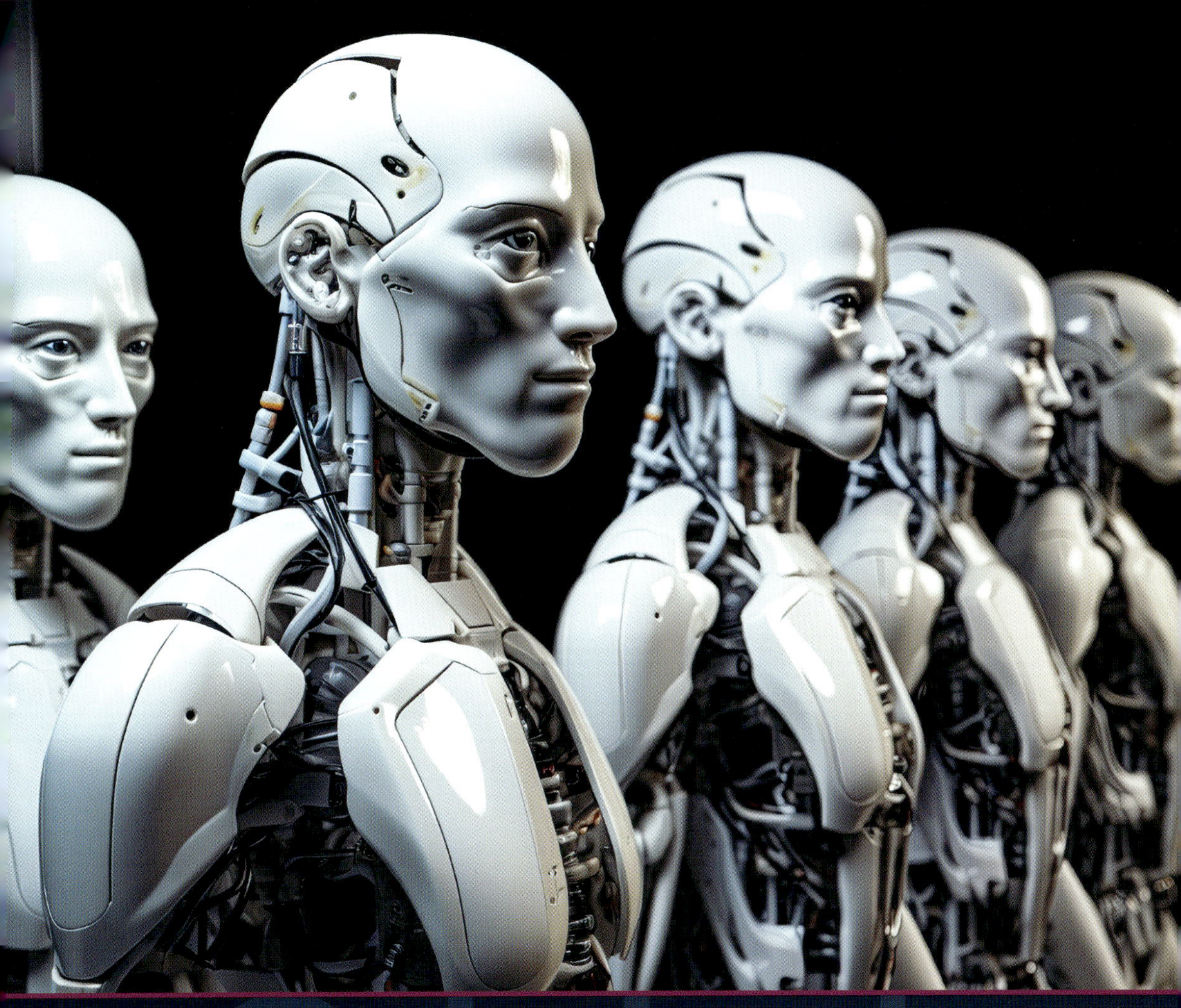

When some people think of AI, they think of robots or droids in movies. AI is getting more advanced every day, but doesn't quite look like the movies, yet.

they learn to solve problems. They can make predictions based on what they already know. They can practice. They can get better at things over time. This is different from other computer programs. Other programs only do things exactly as they are programmed to do them. They cannot learn. They can change only if a programmer rewrites their **code**.

Some people hope to create computers that are as smart as humans. They'd like to bring their favorite science-fiction stories to life. But this is a far-off goal. It might never happen. Today, most people simply want to make AI systems to solve specific problems.

All computer programs require a programmer to write their code. But AI programs can learn and change without more work from a programmer.

A powerful AI system might be created to recommend new products to customers at a store. Another might be designed to recognize different people in photos. These might seem like simple things. You could do them easily without thinking too hard. But getting a computer to do these things isn't so simple. It takes a lot of time and hard work. At first, the computer might not be very good at its job. But the latest AI systems are starting to get very good at certain things. They are almost as good as humans at some jobs. In some cases, they are even better.

AI is not the same thing as human intelligence. It can't exist without human intelligence. All AI systems are created by people. And they are also taught by people. Much like humans, AI systems need to learn and practice to get good at things. This means an AI program is only as good as its training.

THINKING WITHOUT FEELINGS

The human mind is very complex. There is still much we don't know about it. People are not as predictable as computers. It's not always clear how they make decisions. Have you ever "gone with your gut"? This is when you make a decision quickly. You don't stop to think much about it. You simply act. Other times, people do things based on their emotions. They might do something they know is wrong because they want to impress someone. Or they might decide not to do something because it would go against their **morals**. Computers can't make decisions this way. They don't have feelings or morals. They can act only the way people have programmed them to act.

Many people believe that humans have the ability to decide between right and wrong. AI does not have this ability.

Chapter 2

TEACHING MACHINES TO THINK

AI might seem like cutting-edge technology of the future. But the original ideas behind it are not as new as you may think. Computer programmers first started working on AI projects in the 1940s. This wasn't long after the first electronic computers were built.

In 1944, two researchers at the University of Chicago had an idea. They called it a **neural network**. The idea was a computer system that worked like a human brain. It would be made up of smaller parts called **nodes**. These nodes would each do different things. But they would all be connected. This connection would let them trade information back and forth as they "think" and "learn."

Computers of the 1940s were not powerful enough to build neural networks. But today's computers are much more powerful. Neural networks are no longer

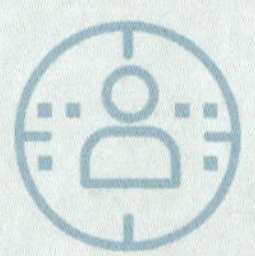

Human brains have neural networks so their brains can synthesize information. AI programs don't have brains, but they have systems that imitate human neural networks.

just an idea. They are powering many of today's newest AI systems.

Multiple neural networks can be connected. This forms something called a deep learning system. Each neural network is a different "layer" of the system. Information can move between the layers. As it does, the layers check each other's work. They improve the system's knowledge. This is how it learns new things. This "deep" stacking of layers on top of each other is also what gives the technology its name.

Deep learning systems are some of the most exciting types of AI in use. They include the technology that lets self-driving cars "see" obstacles on the road using cameras and sensors. They have taught cell phones to understand human speech. They are behind **generative** AI programs like ChatGPT.

Deep learning isn't the only type of AI. It is part of a wider field called machine learning. Other types of machine learning tend to be much simpler than deep learning, though.

AI systems are built around three main things. They use computer **hardware**, data, and **algorithms**. Data can be any kind of **digital** information. It could be photos,

ChatGPT is a generative AI program. It searches and synthesizes the information it has ingested in order to output writing.

videos, or written text. It could be sets of measurements or the results of a science experiment. This is the information the AI system studies to learn.

Algorithms are a series of steps used to complete a specific task. They play an important role in all kinds of computer programming. But they aren't only for computers. A list of steps explaining how to feed and walk your dog could be an algorithm. The ones used in AI systems are much more complex. And many of them can be combined to solve even more complicated problems.

Step 1 for "How to Walk a Dog," would be, "Put on a leash." Step 2 would be, "Go outside." Step 3 would be "Hold the leash while the dog walks." Simple algorithm!

PROCESSING POWER

AI systems require powerful computer hardware. It is only in recent years that technology has become powerful enough for advanced AI.

The hardware that powers many AI systems was not created for that reason. It was created to play video games! Many AI systems use graphics processing units, or GPUs. A GPU is the computer hardware that creates 3D graphics when you play a game. There's one in every gaming console or PC. But researchers discovered that GPUs were also perfect for AI. In an AI system like ChatGPT, thousands of GPUs are connected. They work together like a big brain for the AI.

Chapter 3

PATTERNS AND PREDICTIONS

To train an AI, its creators design algorithms that will look for patterns in data. Each step of the process is broken into smaller and smaller steps. Think of all the steps that might go into recognizing a certain person in a photo. First, the AI needs to be able to tell which parts of the photo are showing human faces. It also needs to know how to tell people apart. Finally, it needs to be able to know what each person looks like in different conditions. For example, people might look different in bright light than at night. They might be making a different facial expression. Their head might be turned to the side.

Each of these steps also includes countless smaller steps. Recognizing someone might seem instant and easy to you. But your brain processes a lot of information every time you do it. The same is true for AI systems.

Many smartphones now have facial recognition software built in. This software must recognize your face in order to unlock log-in and payment information.

AI systems look for patterns in the data they are given. They do not necessarily notice the same things people do. But they can still get humanlike results. This is the power of machine learning. Sometimes even an AI's creators aren't sure exactly *how* the AI figures out certain things. They just know that the AI found patterns in the data.

A trained AI system can use its knowledge to create new data. Imagine an AI that has been trained to identify different kinds of dogs in photos. It knows what makes a golden retriever different from a Chihuahua. Now it can be trained to produce new images of golden retrievers and Chihuahuas.

AI systems are designed by humans. Humans can make mistakes. They also have **biases**. One person's idea of what is "normal" can be very different from another's. These mistakes and biases can affect the way AI systems work. Imagine developers are creating an AI to recognize human speech. They train it using

There are now AI programs that can distinguish between different dog breeds. Then they can create their own dog images of specific breeds.

countless hours of audio recordings of people speaking. They test it. It seems to work great. But then someone who has a different accent tries to use the AI program. It can't recognize that person's speech. This is because the developers didn't think to include recordings of different accents.

This type of problem has popped up a lot in recent years. Some facial recognition systems have a hard time identifying people who aren't of European descent. And some image-creating programs don't recognize cities in other countries. AI's role in our world is growing. AI developers will need to think carefully about making their systems fair and useful for everyone.

ASK AN AI

Let's go straight to the source and ask the AI text generator ChatGPT how it works.

Hi, ChatGPT
Hello! How can I assist you today?

How did you learn to chat?
ChatGPT uses a technology called natural language processing. It's a computer program based on GPT-3.5 that has learned a lot about language by reading a ton of text from the internet. During training, it learned to predict what words come next in a sentence. When you talk to ChatGPT, it uses this knowledge to guess what words or phrases should come next in its responses. It doesn't have real memories or awareness but relies on the context of the current conversation. While it's good for various language tasks, it may have some limitations like biases and occasional mistakes.

Will you get better at chatting over time?
I don't get better with time or user interactions. However, the overall technology might improve if OpenAI releases updated models that benefit from better training methods and new data.

What kinds of data were you trained on?
I learned from many different things on the internet like articles, books, and websites. These cover lots of topics and writing styles to help me understand and talk about different things. But I don't know the specifics about which exact documents were in my training, and I learned up until January 2022.

Thanks for sharing your thoughts!
You're welcome! If you have any more questions or if there's anything else I can help you with, feel free to ask.

(This conversation has been slightly edited and condensed for space.)

Chapter 4

BIG DATA

Think of the way humans learn. Imagine you are taking a math class. What if your textbook was full of errors? What if the teacher told you the wrong way to do everything? It would be very hard for you to learn. You might even end up worse at math than you were when you started. Learning requires good information. It also require good instructions for how to use that information.

The same thing is true when it comes to AI. AI systems are all trained using data. The quality and amount of this data play a big role in how well the AI will perform. People who design AI systems spend a lot of effort gathering and sorting data. They need to make sure they are feeding the right information to their systems. And the more of this information they can find, the better.

Learning from quality data will ensure an AI program produces quality materials.

So where does all this data come from? It would take an enormous amount of time to create it from scratch. Imagine someone is trying to create an AI that recognizes people in video clips. They might need thousands of hours or more of video. It would be very difficult to record all that video footage. Luckily for these researchers, people all around the world are connected to one huge source of all kinds of data—the internet.

The internet has played a huge part in the recent explosion of AI technology. Today's AI developers have access to a nearly limitless amount of data. They can collect video, sound recordings, text, and more. For example, ChatGPT was trained using text gathered from billions of web pages. The words written in digital books, articles, and blogs

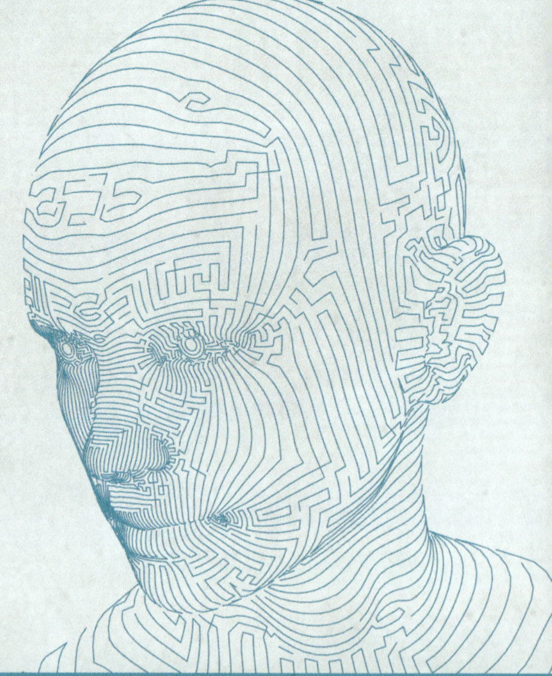

People post questions, thoughts, and ideas all over the internet. Reddit is only one of the millions of sites people post comments on.

helped ChatGPT learn how sentences are put together. It also learned how people write about different subjects in different ways.

Much of this data ends up online because people post it there on purpose. People upload videos to YouTube. They post comments to sites like Reddit and Facebook. They share photos on Instagram. They share original songs on Soundcloud.

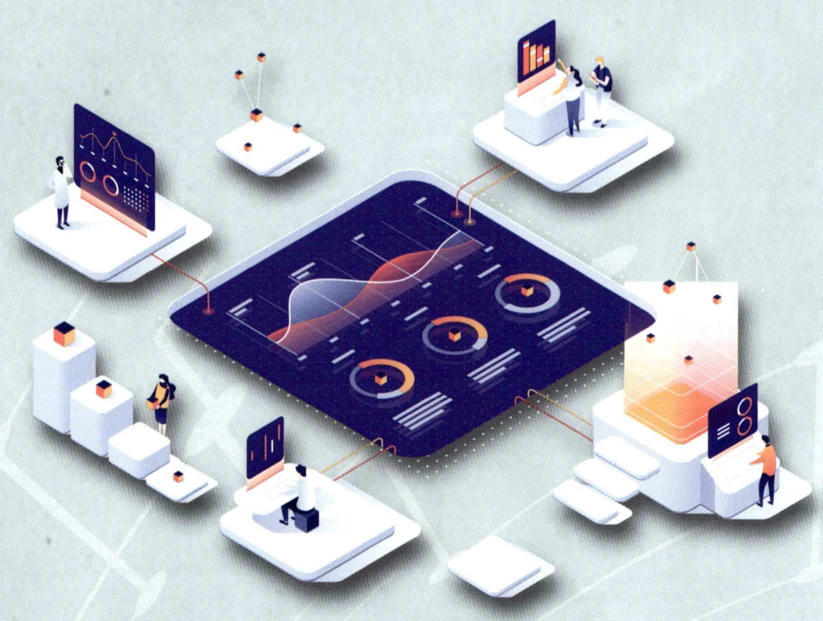

Other data is gathered online in ways you might not even notice. Most websites track everything you look at or click on. Stores keep track of everything you search for or buy. Streaming services track how long you spend watching each video. All of this data might seem pointless. But it can be very helpful for companies—and AI developers.

More data is being created every second of every day. This data will help AI become more and more powerful. It's impossible to predict exactly how AI will affect the future. But one thing is certain. It's going to play a big part in our world.

WHOSE DATA IS IT?

Anything posted online might be used to train an AI. Not everyone is happy about this. Many artists have complained about their work being used to train AI systems. They argue that people should have to give permission for their data to be used this way. Some have tried suing AI companies. They want to keep their data to themselves. This is likely to be a major debate in the coming years. New laws may be passed to change the way AI developers can use data. It's possible that these laws could change the way AI systems are trained in the future.

AI-generated art copying the artist Banksy

ACTIVITY: IT'S YOUR DATA

How much data do you put online every day? The answer might surprise you. Keep a diary one day to see how often you do things online that leave data behind. Some activities to note:

- Visiting a website or clicking a link
- Logging in to social media or any other account online
- Posting something online
- Buying something online
- Using Siri or another voice assistant on your phone
- Using a map app for directions on your phone

These aren't the only activities that track data. They are just some of the major ones. There are ways to prevent some data from being tracked. But the only way to avoid others is to stop using the internet or a smartphone entirely.

FIND OUT MORE

Books

Gitlin, Martin. *The Birth of Modern Tech.* Ann Arbor, MI: Cherry Lake Publishing, 2022.

Gregory, Josh. *Careers in Artificial Intelligence.* Ann Arbor, MI: Cherry Lake Publishing, 2019.

Hulick, Kathryn. *What Is Artificial Intelligence?* Lake Elmo, MN: Focus Readers, 2020.

Kulz, George Anthony. *Artificial Intelligence in the Real World.* Lake Elmo, MN: Focus Readers, 2020.

On the Web

Search these online sources with an adult:

"Algorithm facts for kids." Kiddle.

"Artificial Intelligence." Britannica for Kids.

"ChatGPT." OpenAI.

"Machine learning facts for kids." Kiddle.

"What is Artificial Intelligence?" Duke University on YouTube.

"What is artificial intelligence (AI)?" IBM.

GLOSSARY

algorithms (AL-guh-rih-thuhmz)
series of steps that can be followed to achieve a certain goal

biases (BYE-uhs-iz)
unfair preferences for certain things

code (KOHD)
instructions written in computer programming language

data (DAY-tuh)
information used to create, process, or support something

digital (DIH-juh-tuhl)
relating to electronic computers

generative (JEH-nuh-ruh-tiv)
able to create something

hardware (HARD-wair)
physical computer equipment

morals (MOHR-uhlz)
beliefs that guide peoples' actions

neural network (NER-uhl NET-werk)
computer system built to work like human brains

nodes (NOHDZ)
individual pieces of a larger connected network

INDEX